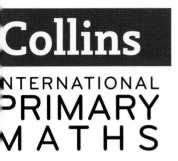

Collins

INTERNATIONAL
PRIMARY
MATHS

Student's Book 1

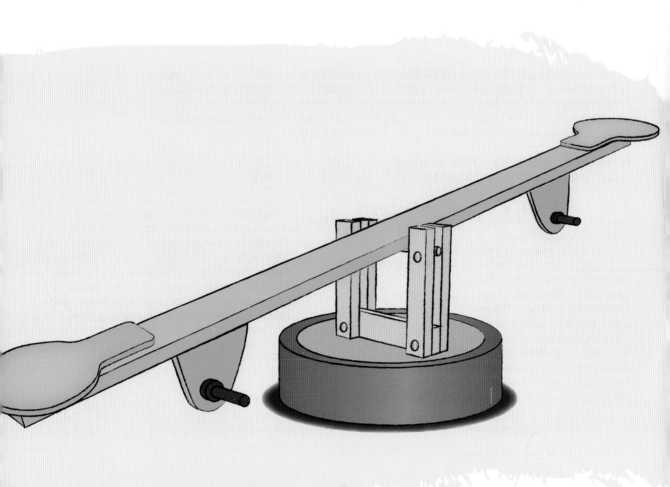

William Collins' dream of knowledge for all began with the publication of his first book in 1819. A self-educated mill worker, he not only enriched millions of lives, but also founded a flourishing publishing house. Today, staying true to this spirit, Collins books are packed with inspiration, innovation and practical expertise. They place you at the centre of a world of possibility and give you exactly what you need to explore it.

Collins. Freedom to teach.

Published by Collins
An imprint of HarperCollins*Publishers*
The News Building
1 London Bridge Street
London
SE1 9GF

HarperCollins*Publishers*
Macken House, 39/40
Mayor Street Upper, Dublin 1,
D01 C9W8, Ireland

Browse the complete Collins catalogue at
www.collins.co.uk

British Library Cataloguing-in-Publication Data
A catalogue record for this publication is available from the British Library.

Author: Lisa Jarmin
Series editor: Peter Clarke
Publisher: Elaine Higgleton
Product developer: Holly Woolnough
Project manager: Mike Harman (Life Lines Editorial Services)
Development editor: Joan Miller
Copyeditor: Tanya Solomons
Proofreader: Catherine Dakin
Cover designer: Gordon MacGilp
Cover illustrator: Ann Paganuzzi
Typesetter: Ken Vail Graphic Design
Illustrators: Ann Paganuzzi and QBS Learning
Production controller: Lyndsey Rogers
Printed and bound in India by Replika Press Pvt. Ltd.

With thanks to the following teachers and schools for reviewing materials in development: Calcutta International School; Hawar International School; Melissa Brobst, International School of Budapest; Rafaella Alexandrou, Pascal Primary Lefkosia; Maria Biglikoudi, Georgia Keravnou, Sotiria Leonidou and Niki Tzorzis, Pascal Primary School Lemessos; Taman Rama Intercultural School, Bali.

The publishers gratefully acknowledge the permission granted to reproduce the copyright material in this book. Every effort has been made to trace copyright holders and to obtain their permission for the use of copyright material. The publishers will gladly receive any information enabling them to rectify any error or omission at the first opportunity.
Cambridge International copyright material in this publication is reproduced under licence and remains the intellectual property of Cambridge Assessment International Education

Contents

Number

Geometry and Measure

Statistics and Probability

How to use this book

This book is used at the start of a lesson when your teacher is explaining the mathematical ideas to the class.

- An **objective** explains what you should know, or be able to do, by the end of the lesson.

Key words

- The **key words** to use during the lesson are shown. It's important that you understand what each of these words mean.

Let's learn

This part of the Student's Book page **teaches** you the main mathematical ideas of the lesson. It might include pictures or diagrams to help you **learn**.

Guided practice

Guided practice helps you to answer the questions in the Workbook. Your teacher will talk to you about this question so that you can work by yourself on the Workbook page.

HINT

Use the page in the Student's Book to help you answer the questions on the Workbook pages.

 Thinking and Working Mathematically (TWM) involves thinking about the mathematics you are doing to gain a deeper understanding of an idea, and to make connections with other ideas. The TWM Star at the back of this book explains the 8 different ways of working that make up TWM.

At the back of the book

Number

Lesson 1: **Counting objects**

- Count up to 10 objects

Let's learn

Guided practice
Count the sweets.

5

Lesson 2: **Counting on and back in ones**

• Count forwards and backwards in ones to 10

Number

Let's learn

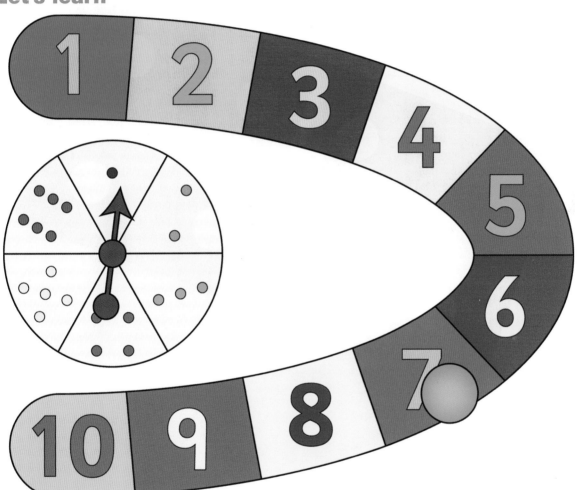

Guided practice

Count forwards in ones. Which number goes in the star?

1		3		5		☆7	8		10

7

Lesson 3: **Sequences of objects**

Key words
- **number**
- **pattern**

• Recognise numbers as patterns

Let's learn

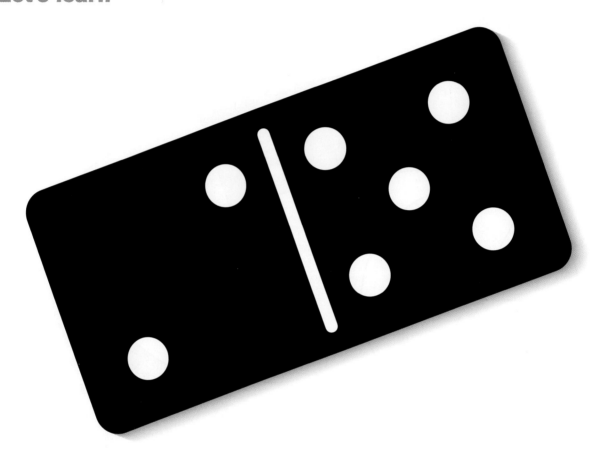

Guided practice
Tick the number that matches.

 3 ☐ 5 ☐ 4 ✔

Lesson 4: **Estimating to 10**

- Estimate how many objects there are and check this by counting

Key words
- estimate
- guess
- more than
- less than

Number

Let's learn

Guided practice

Estimate, then count.

Estimate 5   Count 6

Lesson 1: **Estimating to 20**

• Estimate amounts of objects to 20

Key words
• **estimate**
• **count**

Let's learn

Guided practice

Estimate, then count.

Estimate | 15 |

Amount | 13 |

Lesson 2: **Counting in twos**

- Count on in twos from any number to 20
- Notice and describe number patterns

Number

Let's learn

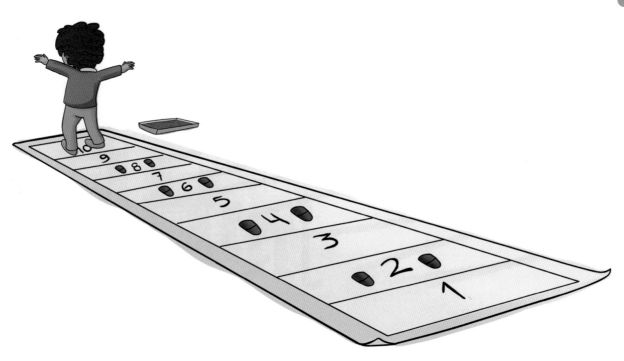

Guided practice

Use the number line to help you count on in twos.

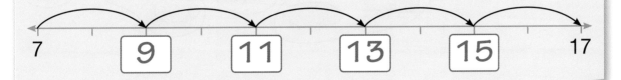

Lesson 3: **Odd and even numbers**

Key words
- **odd**
- **even**

- Recognise odd and even numbers to 20

Number

Let's learn

Guided practice

How many pebbles? | 14 |
Is the number odd
or even? | *even* |

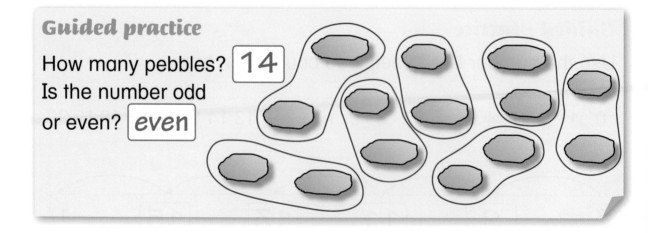

12

Lesson 4: **Counting in tens**

- Count on and back in tens

Key words
- tens
- count
- on
- back
- more
- less

Number

Let's learn

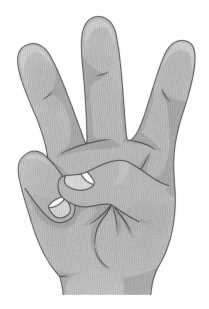

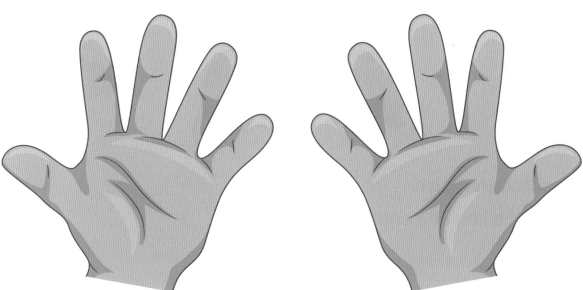

Guided practice

Count on or back in tens to find the answers.

10 **more** than 8 is: | 18 | 10 **less** than 13 is: | 3 |

Lesson 1: **Counting to 10**

Key words
- **number**
- **count**
- **symbols**

• Count to 10

Let's learn

Guided practice

Trace the line and count on.

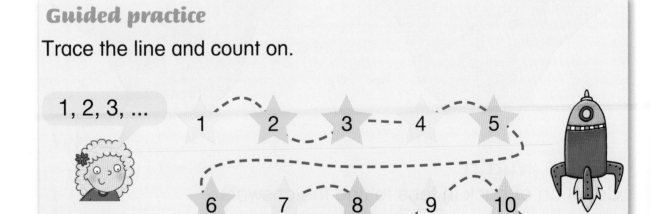

1, 2, 3, ...

Lesson 2: **Reading numbers to 10**

Key words
• **number**
• **digit**

• Read the numbers 1 to 10

Number

Let's learn

Guided practice

Count and then draw a line from each group to the matching number.

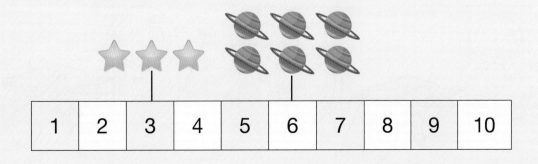

Number

Lesson 3: **Writing numbers to 10**

• Write the numbers 1 to 10

Let's learn

Shopping list
8 apples
2 bananas
6 tomatoes

Guided practice

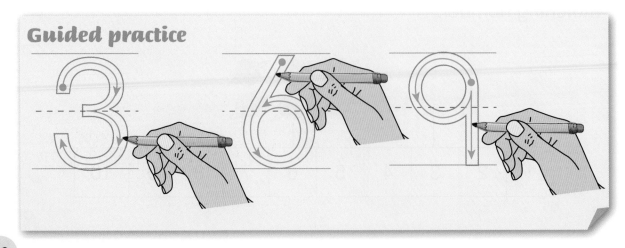

Lesson 4: **How many?**

• Count up to 10 objects

Key word
• **count**

Number

Let's learn

Guided practice

How many fir cones?   5

Circle each one as you count it.

Lesson 1: **Counting to 20**

Key words
• count
• forwards
• backwards

• Count forwards and backwards from any given number to 20

Let's learn

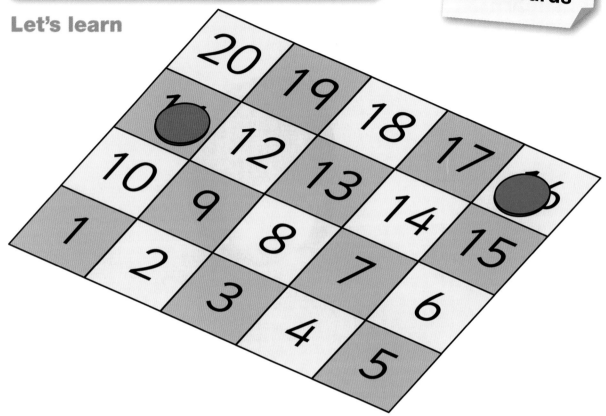

Guided practice

Trace the line and count on.

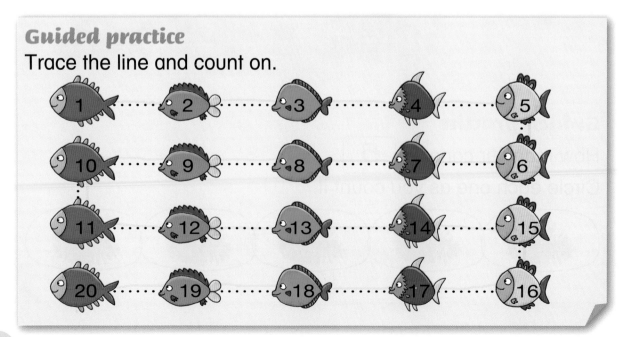

Lesson 2: **Reading numbers to 20**

Key words
- **number**
- **numeral**

• Read numbers to 20

Let's learn

thirteen   ten   sixteen   four   twenty

Guided practice

Count the number of cubes or counters.
Draw lines to join a word and a numeral to each set.

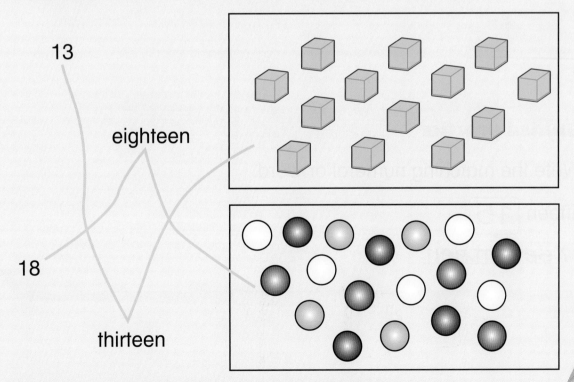

13

eighteen

18

thirteen

Number

Lesson 3: **Writing numbers to 20**

- Write numerals and number words to 20

Let's learn

Flat 17
Green Street
North Town
NT5 5JR

Guided practice

Write the matching numeral or word.

fifteen 15

17 _seventeen_

Lesson 4: **Counting and labelling objects to 20**

Key words
• count
• number
• how many

Number

- Count up to 20 objects accurately
- Label amounts to 20

Let's learn

19

Guided practice

Count the butterflies. Write how many there are as a numeral and a word.

18 _eighteen_

Lesson 1: **Combining sets**

- Combine two sets to find how many altogether

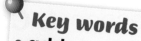

Key words
- **add**
- **count**
- **altogether**

Let's learn

Guided practice

How many altogether? | 7 |

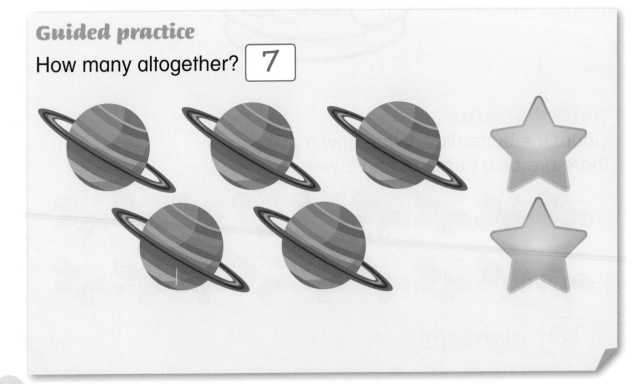

Lesson 2: **Part–whole diagrams**

Number

• Use part–whole diagrams to combine sets of objects

Let's learn

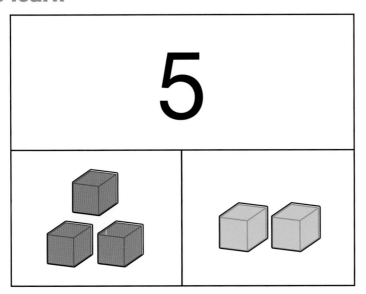

5

Guided practice

How many altogether?

6

Number

Lesson 3: **Writing addition number sentences**

- Use + and = in an addition number sentence
- Use groups of objects to make an addition number sentence

Key words
- add
- plus
- count
- altogether
- augend
- addend
- sum
- total
- number sentence

Let's learn

+

=

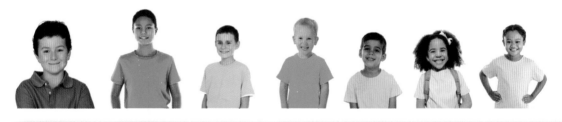

Guided practice

Complete the number sentence to match the group of objects.

3 + 2 = 5

Lesson 4: **Using sets of objects to solve additions**

Key words
- add
- count
- altogether
- part–whole diagram
- augend
- addend
- sum
- total
- number sentence

Number

- Solve additions by combining sets of objects

Let's learn

$5 + 2 =$

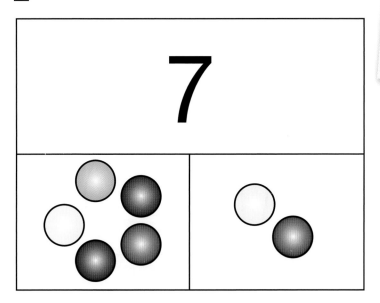

Guided practice

Use the part–whole diagram to find the answer.

$5 + 1 = \boxed{6}$

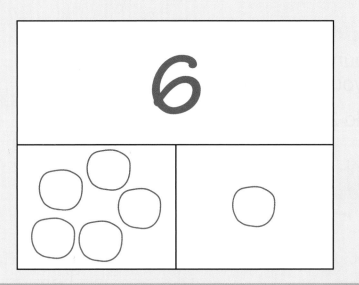

Number

Lesson 1: **Adding more**

- Add two amounts by counting on with objects

Key words
- add
- count on
- more
- sum
- total

Let's learn

Guided practice

Count on from the starting amount.
If you need to, use counters to help.

Add 2 more.

7

Lesson 2: **Adding more to solve additions**

- Solve additions by counting on more objects

Key words
- add
- count on
- augend
- addend
- sum
- total
- number sentence

Let's learn

3 + 2 =

Guided practice

Draw the extra shapes to count on.

6 + 3 = 9

| 1 | 2 | 3 |

| 4 | 5 | 6 |

Number

Lesson 3: **Adding more with a number track**

- Use a number track to count on to solve additions

Key words
- add
- count on
- augend
- addend
- sum
- total
- number sentence

Let's learn

5 + 2 =

2 more jumps

Guided practice

Count on along the number track from the larger number to find the total.

 1 2 3 4 5 6 7 8 ⑨ 10

6 + 3 = 9

Lesson 4: **Adding more with a number line**

• Use a number line to count on to solve additions

Let's learn

Key words
• add
• count on
• augend
• addend
• sum
• total
• number sentence

Number

3 jumps

Guided practice

Draw jumps on the number line to count on and draw a ring around the total. Remember to start on the larger number.

6 + 2 = **8**

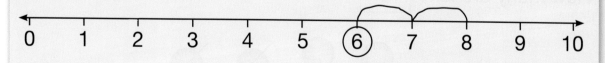

Number

Lesson 1: **Taking away objects**

- Take objects from a set and re-count to find how many are left

Key words
- **take away**
- **subtract**
- **group**

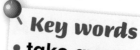

Let's learn

Guided practice

Count out 8 counters. Now take away 2.

How many are left? `6`

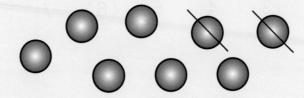

Lesson 2: **Taking away to solve subtractions**

- Solve subtractions by taking away objects from a group

Number

Let's learn

$6 - 4 = 2$

Guided practice

Take away the number shown.

   **3** How many are left? 2

Lesson 3: **Taking away with part–whole diagrams**

Key word
• part–whole diagram

• Use a part–whole diagram to take away

Let's learn

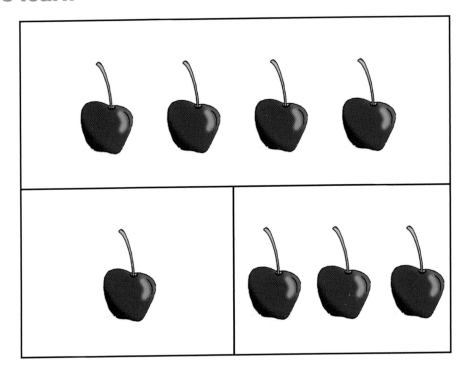

Guided practice

Draw the counters to complete the part–whole diagrams.

Take away 4.

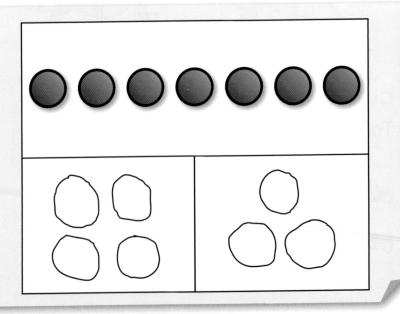

Lesson 4: **Solving subtractions with part–whole diagrams**

Number

• Use a part–whole diagram to solve subtractions

Let's learn

Leila had 7 sweets. She gave 2 sweets to her brother. How many sweets did Leila have left?

$7 - 2 =$

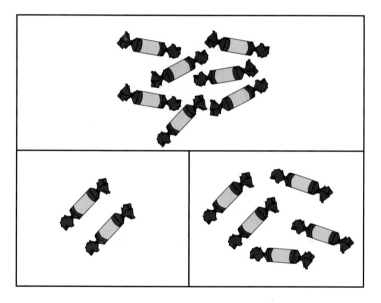

Guided practice

Use your part–whole diagram to find the answer to the subtraction.

$8 - 5 =$ 3

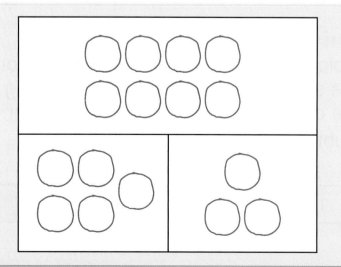

Lesson 1: **Counting back in ones to subtract**

* Count back in ones to subtract an amount

Let's learn

10, 9, 8,
7, 6, 5, 4,
3, 2, 1, 0

Guided practice

Colour in the number of shapes that you are subtracting. Then count back along the coloured shapes to find the answer.

Subtract 2.

| 1 | 2 | 3 | 4 | 5 | 6 | 4 |

Lesson 2: **Subtracting on a number track**

Key words

- **count back**
- **subtract**
- **number track**
- **subtrahend**
- **minuend**
- **equals**
- **number sentence**

- Solve subtractions by counting back on a number track

Let's learn

6 – 2 =

Guided practice

Solve the subtraction by counting back.

0 1 2 3 4 5 6 7 8 9 10

7 – 4 = 3

Number

Lesson 3: **Subtracting on a number line**

- Solve subtractions by counting back on a number line

Let's learn

$8 - 3 = 5$

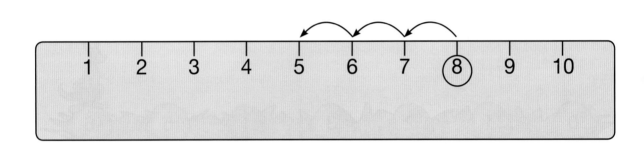

Guided practice

Use the number line to solve the subtraction.

$8 - 5 = \boxed{3}$

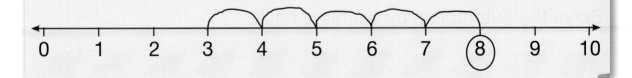

Lesson 4: **Counting back to solve subtractions**

- Use mental strategies to count back to solve subtractions

Number

Let's learn

$$7 - 3 = 4$$

Guided practice

Count back in your head to solve the subtraction. Then use the number line to check your answer.

$$10 - 4 = \boxed{6}$$

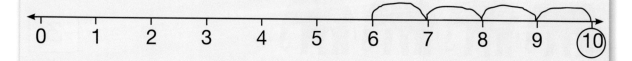

Lesson 1: **Finding the difference**

• Find the difference
between two amounts

Let's learn

5 more

Guided practice
Circle the extra animals in the top line.
Count them to find the difference.

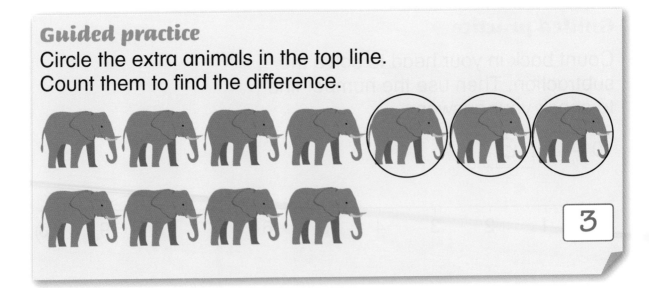

3

Number

Lesson 2: **Subtracting by finding the difference**

• Solve subtractions by finding the difference

Let's learn

$8 - 3 =$

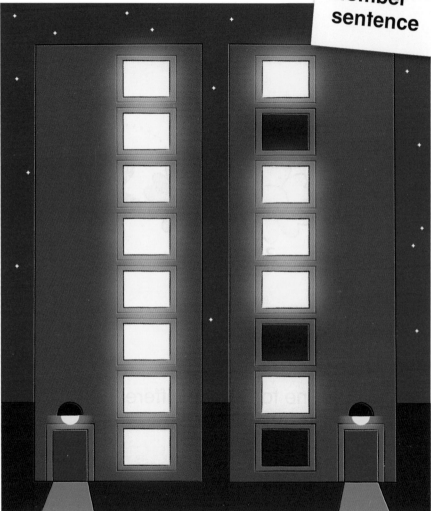

Guided practice

Draw counters to find the difference.

$5 - 4 = \boxed{1}$

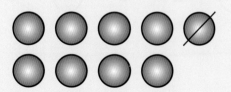

Lesson 3: **Finding the difference on a number line**

- Find the difference between two numbers on a number line

Key words
- difference
- number line

Let's learn

Guided practice

Use the number line to find the difference between the pair of numbers.

4 and 1   3

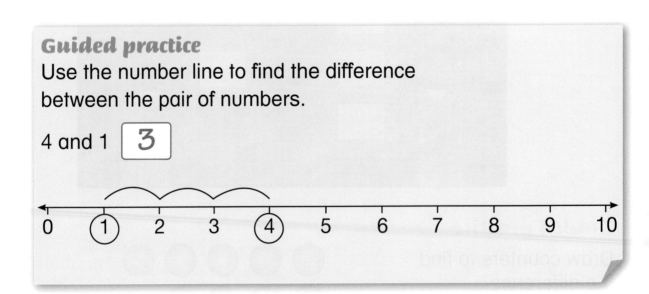

Lesson 4: **Subtraction as difference on a number line**

Number

Key words
- difference
- subtract
- minuend
- subtrahend
- equals
- number sentence

- Solve subtractions by finding the difference on a number line

Let's learn

$$8 - 3 =$$

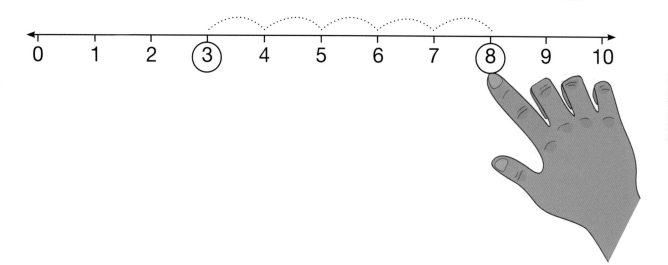

Guided practice

Use the number line to find difference.

$$4 - 1 = \boxed{3}$$

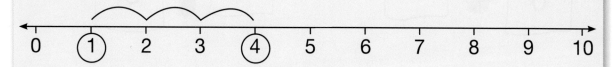

Lesson 1: **Making 10**

- Find pairs of numbers that total 10

Let's learn

Guided practice

Draw sweets in the empty bag to make 10 sweets altogether.

Write how many sweets are in each bag.

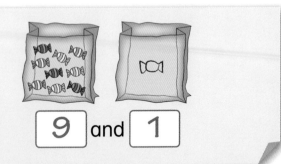

9 and 1

Lesson 2: **Addition and subtraction facts for 10**

- Find addition and subtraction facts for 10

Number

Let's learn

$8 + 2 = 10$

$2 + 8 = 10$

$10 - 2 = 8$

$10 - 8 = 2$

Guided practice

Write one addition and one subtraction number sentence.

10	
8	2

$8 + 2 = 10$

$10 - 2 = 8$

Lesson 3: **Making numbers to 10**

- Find pairs of numbers that make totals to 10

Let's learn

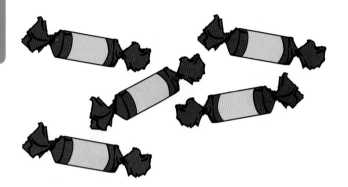

Guided practice

Use cubes to help you find pairs of numbers that total 4.

(4)

4 and [0]   0 and [4]

3 and [1]   1 and [3]

[2] and [2]

Lesson 4: **Addition and subtraction facts to 10**

- Find addition and subtraction facts to 10

Number

Let's learn

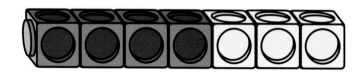

$4 + 3 = 7$

$3 + 4 = 7$

$7 - 3 = 4$

$7 - 4 = 3$

7	
4	3

Guided practice

Write two additions and two subtractions for the complement of 6.

6	
5	1

$\boxed{5} + \boxed{1} = 6$

$\boxed{1} + \boxed{5} = 6$

$6 - \boxed{5} = \boxed{1}$

$6 - \boxed{1} = \boxed{5}$

Lesson 1: **Estimating an answer**

• Estimate the answer to an addition or subtraction to 10

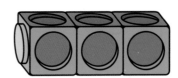

Key words
• addition
• subtraction
• estimate

Let's learn

$5 + 3 =$

I estimate less than 10.

My estimate is more than 5.

I think the answer is about 7.

Guided practice

$5 + 2 =$ My estimate: $\boxed{8}$

Answer: $\boxed{7}$

Lesson 2: **Choosing how to solve an addition**

• Use different strategies to solve additions

Key words
• add
• augend
• addend
• equals
• number sentence
• count on
• combine sets

Let's learn

$6 + 4 =$

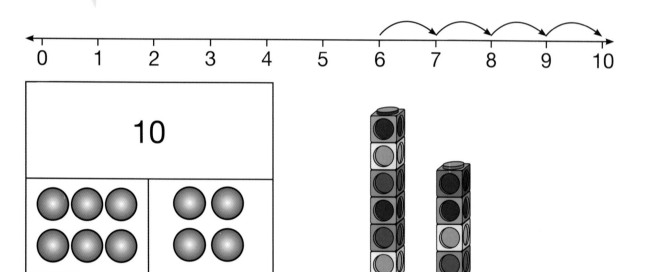

Guided practice

Choose a method or piece of equipment to solve the addition.

$3 + 2 = \boxed{5}$

3... 4, 5.

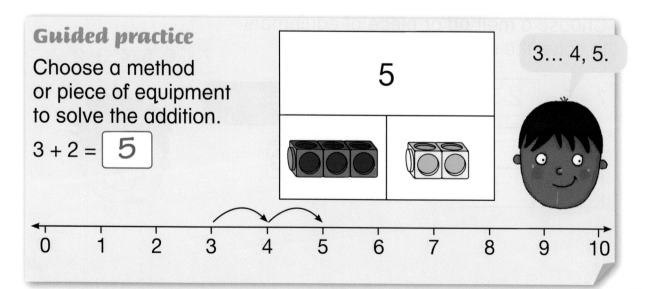

Number

Lesson 3: **Choosing how to solve a subtraction**

- Use different strategies to solve subtractions

Let's learn

$7 - 4 =$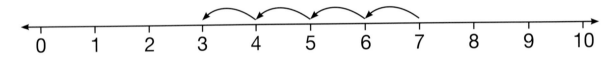

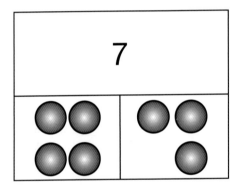

7

Guided practice

Choose a method or piece of equipment to solve the subtraction.

7... 6, 5, 4, 3, 2.

$7 - 5 = \boxed{2}$

7	
5	2

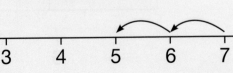

Lesson 4: **Addition and subtraction in real life**

Key words
- addition
- subtraction

- Relate addition and subtraction to real-life situations
- Use a chosen strategy to solve additions and subtractions

Let's learn

Guided practice

Write the number sentence.

Fin makes a tower of 7 bricks. Rashi knocks 2 bricks off. How many bricks are left in Fin's tower?

$$7 - 2 = 5$$

Lesson 1: **Addition facts to 20**

- Add 1-digit and 2-digit numbers to 20 by counting on

Number

Let's learn

Guided practice

Use the number line to count on.

0 1 2 3 4 5 6 7 8 9 10 11 12 13 14 15 16 17 18 19 20

$13 + 3 = \boxed{16}$

50

Lesson 2: **Making 10 and adding more**

Number

Key words
- complements
- ten
- add
- more

- Use number pairs that total 10 to add

Let's learn

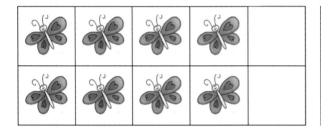

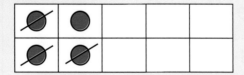

It's easy to add a number to 10.

Guided practice

First make 10, then add what's left over.

$$7 + 4 = 11$$

Number

Lesson 3: **Near doubles**

- Solve additions by finding near doubles

Key words
- **add**
- **double**

Let's learn

Guided practice
Use doubling to work out the answer.

3 + 4

Double ⬚3 = ⬚6

⬚6 ⊕ ⬚1 = ⬚7

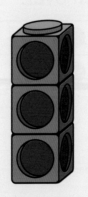
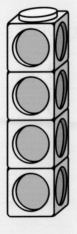

Lesson 4: **Addition and estimation to 20**

Key words
- add
- estimate
- double

- Use different strategies to solve additions
- Estimate an answer for an addition

Let's learn

$8 + 4 =$

Find the larger number first then count on.

Make 10 and add more.

Use doubles.

Guided practice

Estimate, then choose a strategy to solve the addition.

There are 5 butterflies on the bush. There are 4 butterflies on the flowers. How many butterflies are there altogether?

Estimate: = 8   5 + 4 = 9

I will use double 4 to work out the answer.

Lesson 1: **Subtraction facts to 20**

- Subtract a 1-digit number from a 2-digit number to 20

Key words
- **tens**
- **ones**
- **subtract**
- **count back**

Let's learn

$17 - 3 =$

I know that $7 - 3 = 4$ so I can use this to help me.

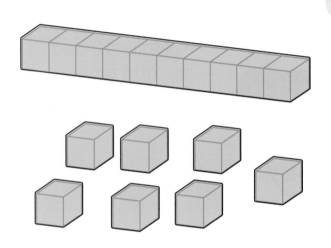

Guided practice

Use what you know to solve the subtraction.
Then use the number line to check your answer.

0 1 2 3 4 5 6 7 8 9 10 11 12 13 14 15 16 17 18 19 20

$17 - 2 =$?

$7 - 2 =$ 5

So $17 - 2 =$ 15

Lesson 2: **Number facts to 20 – part–whole diagrams**

- Find number facts to 20 by using a part–whole diagram

Key words
- addition
- subtraction
- add
- subtract
- equals
- part–whole diagram

Number

Let's learn

16

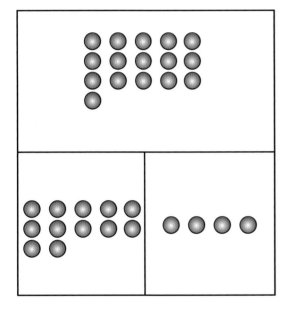
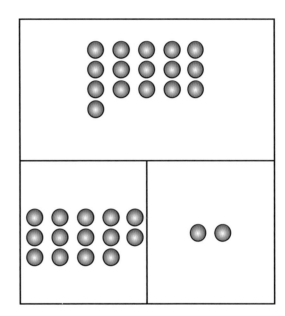

Guided practice

Draw dots to find two different ways of making 12.

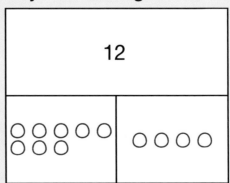

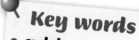

Lesson 3: **Number families**

Number

- Find addition and subtraction statements for number families to 20

Key words
- **add**
- **subtract**
- **equals**
- **number families**

Let's learn

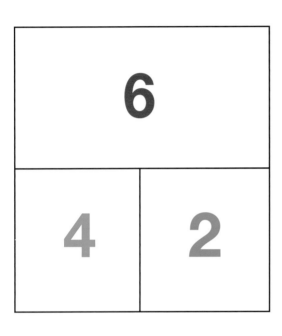

$$4 + 2 = 6$$

$$2 + 4 = 6$$

$$6 - 4 = 2$$

$$6 - 2 = 4$$

Guided practice

Write additions and subtractions to match the number family.

8
5	3

$5 + 3 = 8$ $3 + 5 = 8$

$8 - 5 = 3$ $8 - 3 = 5$

Lesson 4: **Equal statements**

- Find equal statements for numbers to 20

Key words
- add
- subtract
- equals

Number

Let's learn

7 + 2

10 − 1

Guided practice

Write equal statements to match the number.

8    $5 + 3 = 10 - 2$

Lesson 1: **Doubling amounts to 5**

Key word
• **double**

• Find and make doubles for amounts to 5

Let's learn

Guided practice

Draw the same number of spots on the ladybird to make double the amount.

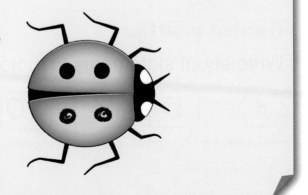

Double 2 = ☐4

Lesson 2: **Doubling amounts to 10**

Key word
• **double**

• Find and make doubles for amounts to 10

Let's learn

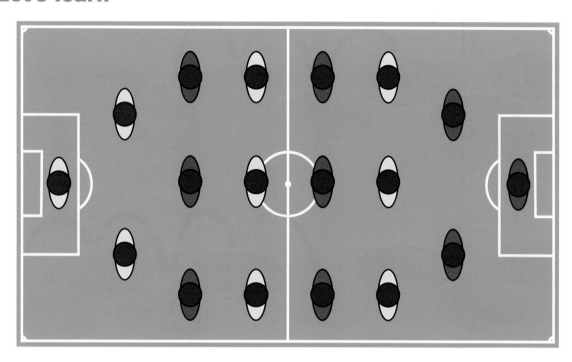

Guided practice

Draw the same amount to make double.

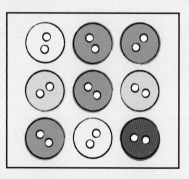

Double 9 = ☐ 18

Lesson 3: **Doubling on a number line**

• Find doubles to 10 on a number line

Let's learn

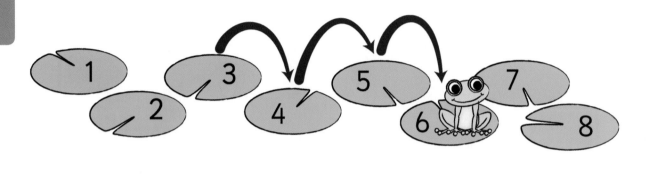

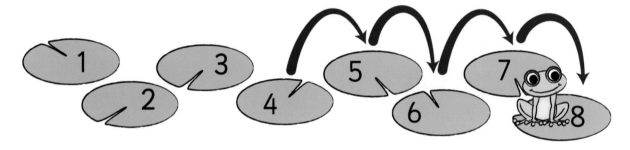

Guided practice

Count along the number line to double the number.

Double 8 = $\boxed{16}$

0 1 2 3 4 5 6 7 8 9 10 11 12 13 14 15 16 17 18 19 20

Number

Lesson 4: **Doubling facts to 10**

Key word
• **double**

• Recall and use doubling facts to 10

Let's learn

1 5 6 4

IN

Double

OUT

→ 2
→ 10
→ 12
→ ?

Guided practice

Double the number. Write it in the box.

Double 4 = $\boxed{8}$

Number

Lesson 1: **What is money?**

- Know what money is used for
- Recognise features of the money we use

Let's learn

Guided practice

Circle any coins that you recognise.
If you don't recognise any, draw a
coin that you know.

Lesson 2: **Recognising coins**

• Recognise the coins we use

Key words
• **money**
• **coins**

Let's learn

Guided practice

Draw three different coins that you use.
Think about their colour and shape.
What markings do they have?

Lesson 3: **Recognising notes**

Key words
• **money**
• **notes**

• Recognise the notes we use

Let's learn

Guided practice

Draw a banknote that you use. Think about the colour and what's shown on the note.

Lesson 4: **Sorting coins and notes**

Key words
• **money**
• **coins**
• **notes**
• **sort**

• Sort the money we use in different ways

Number

Let's learn

Guided practice

Choose how you would like to sort the coins.
Write a heading in each box to show your
sorting rule then draw the coins.

Round	Other shapes

Lesson 1: **Zero**

* Understand that zero means nothing

Let's learn

Guided practice

Draw 2 more :

Draw 0 more :

How many now? 5

How many now? 3

Lesson 2: **Comparing numbers to 10**

Key words
- compare
- more
- less
- greater
- smaller

Number

- Compare sets of objects and say which has more or less
- Compare numbers to 10

Let's learn

Abdul

Karim

Guided practice

Draw dots in the box to make this correct.

more

less

Number

Lesson 3: **Ordering numbers to 10**

• Order numbers to 10
• Give a number between two numbers

Key words
• order
• smallest
• greatest
• between

Let's learn

Guided practice

Fill in the missing number.

6, ⬚**7**⬚ , 8

Order the numbers, starting with the smallest.

8, 1, 4   ⬚**1**⬚ , ⬚**4**⬚ , ⬚**8**⬚

68

Lesson 4: **Ordinal numbers**

• Use ordinal numbers to show position

Let's learn

Key words
- first – 1st
- second – 2nd
- third – 3rd
- fourth – 4th
- fifth – 5th
- sixth – 6th
- seventh – 7th
- eight – 8th
- ninth – 9th
- tenth – 10th
- last

Number

Guided practice
Write the ordinal numbers on the crocodiles, in order.

 1st

 2nd

 3rd

Lesson 1: **Partitioning numbers into tens and ones**

- Partition numbers from 10 to 20 into tens and ones

Key words
- partition
- decompose
- tens
- ones

Let's learn

13 is made of 1 ten and 3 ones.

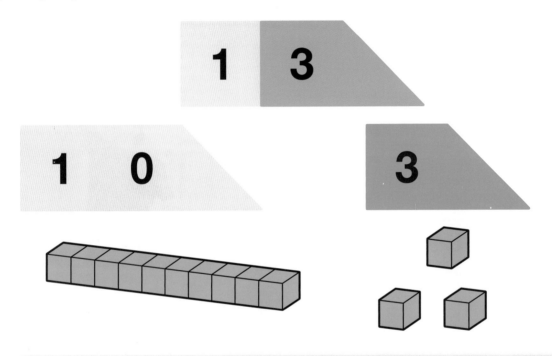

Guided practice

Draw tens and ones blocks to match the number.

17

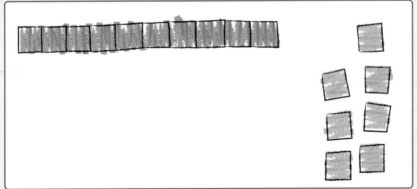

Lesson 2: **Combining tens and ones**

Number

• Make numbers from 10 to 20 from tens and ones

Let's learn

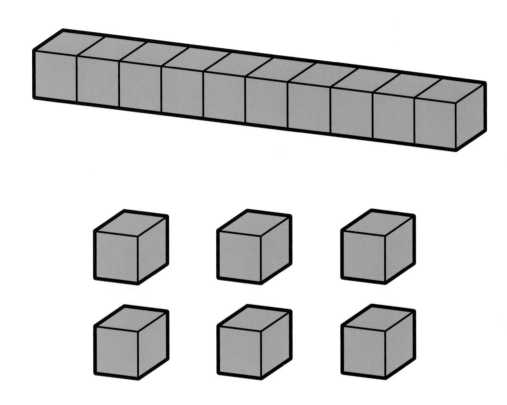

Guided practice

Write the number shown.

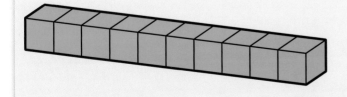   12

Lesson 3: **Representing numbers in different ways**

Number

Key words
- **numbers**
- **tens**
- **ones**
- **regrouping**

- Express a number to 20 in different ways

Let's learn

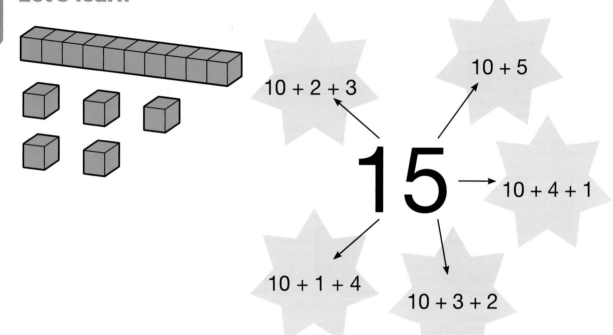

10 + 2 + 3

10 + 5

15

10 + 4 + 1

10 + 1 + 4

10 + 3 + 2

Guided practice

How many ways can you think of to regroup 13?

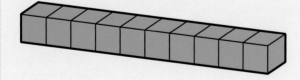

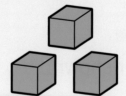

10 + $\boxed{3}$ = 13

10 + $\boxed{2}$ + $\boxed{1}$ = 13

10 + $\boxed{1}$ + $\boxed{2}$ = 13

Lesson 4: **Comparing and ordering numbers to 20**

- Compare and order numbers to 20

Key words
- compare
- order
- more / most
- greater / greatest
- less / least
- smaller / smallest

Number

Let's learn

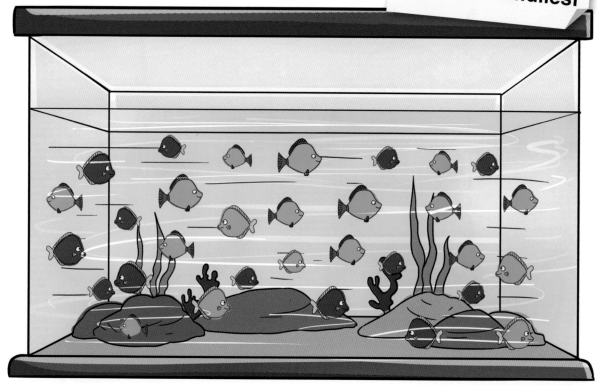

Guided practice

Use the number line to order these numbers, smallest to greatest.

0 1 2 3 4 5 6 7 8 9 10 11 12 13 14 15 16 17 18 19 20

 17   6   18

 6   17   18

Number

Lesson 1: **Halving objects**

- Find half of an object
- Recognise objects that are in halves

Let's learn

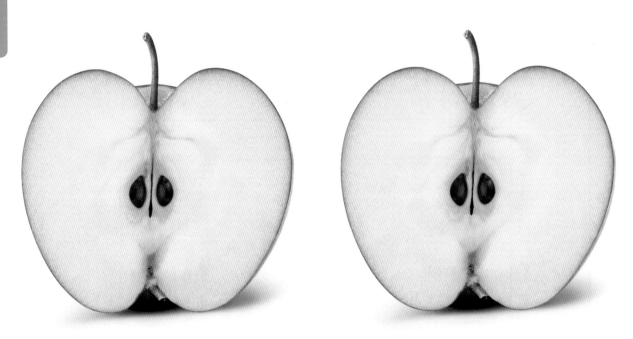

Guided practice

Circle the food that is in halves.

Lesson 2: **Halves of shapes**

- Find half of a shape
- Recognise shapes that are in halves

Number

Let's learn

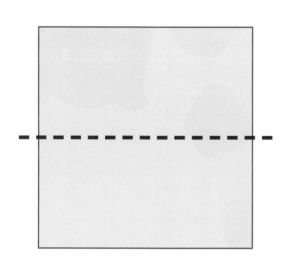

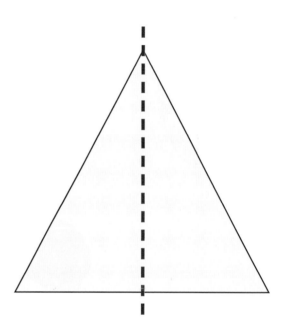

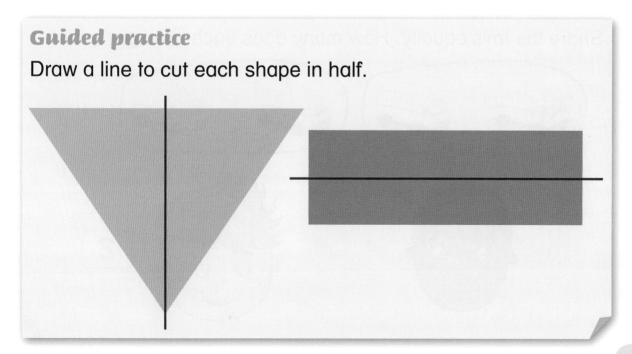

Guided practice

Draw a line to cut each shape in half.

75

Number

Lesson 3: **Halves of amounts (1)**

- Find half of an amount

Let's learn

Guided practice

Share the toys equally. How many does each child get?

2

Lesson 4: **Halves of amounts (2)**

- Find half of an amount

Key words
- **whole**
- **half**

Number

Let's learn

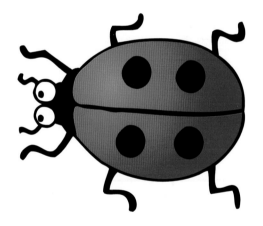

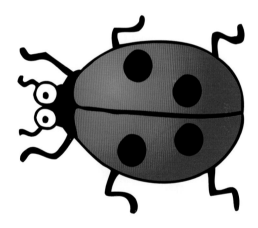

Guided practice

Share the spots equally.

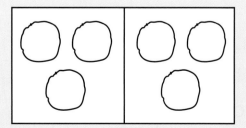

Number

Lesson 1: **Halving numbers to 10**

- Find half of a number to 10
- Use half as an operator

Let's learn

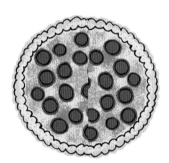

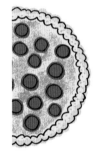

$$\boxed{10}\ \boxed{5}$$

Guided practice

Find half of the number.

Half of 6 ⟶ $\boxed{3}$

Lesson 2: **Halving numbers to 20**

Key words
- whole
- half
- halves
- operator

- Find half of a number to 20
- Use half as an operator

Let's learn

$\dfrac{1}{2}$

$\dfrac{1}{2}$

$\dfrac{1}{2}$

Guided practice

Find half of the number.

 $\dfrac{1}{2}$ of  = 4

Number

Lesson 3: **Combining halves (1)**

• Put halves together to make wholes

Let's learn

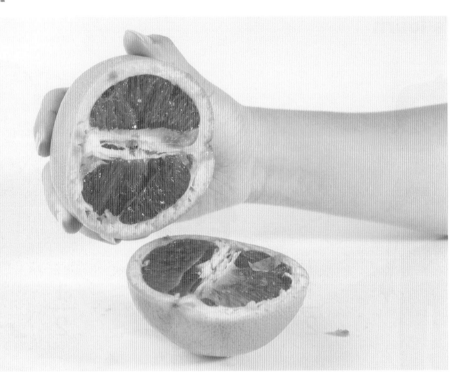

Guided practice

Count the halves.

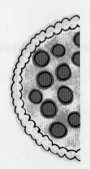 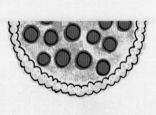 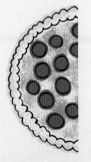 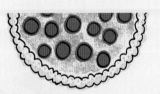

There are 2 whole pizzas.

Lesson 4: **Combining halves (2)**

• Combine halves of an amount back into the whole amount

Key words
• whole
• half

Number

Let's learn

Guided practice

Half of the amount is shown. Draw the other half.
Count all the dots to find the whole amount.

10

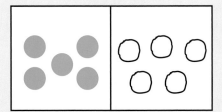

Unit **20** Time

Workbook page 82

Lesson 1: **Days of the week**

- Know the days of the week
- Understand that there are weekdays and weekend days

Geometry and Measure

Let's learn

Monday

Sunday   Tuesday

Saturday   Wednesday

Friday   Thursday

Guided practice

Write the day that comes next.

Monday, Tuesday,   Wednesday

Lesson 2: **Months of the year**

- Know the months of the year
- Know some familiar events that happen in each month

Key words
- year
- month
- week

Let's learn

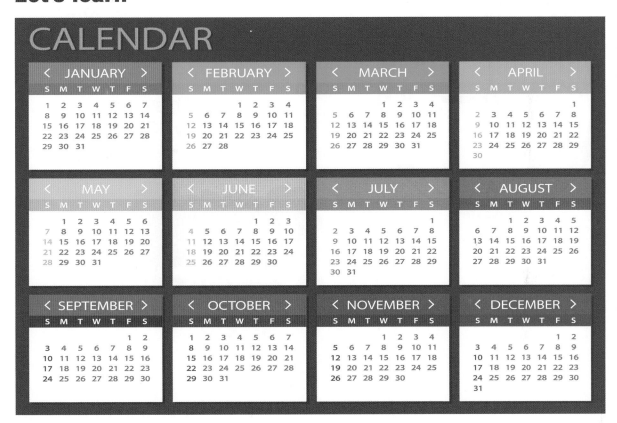

CALENDAR

JANUARY — S M T W T F S — 1 2 3 4 5 6 7 / 8 9 10 11 12 13 14 / 15 16 17 18 19 20 21 / 22 23 24 25 26 27 28 / 29 30 31

FEBRUARY — S M T W T F S — 1 2 3 4 / 5 6 7 8 9 10 11 / 12 13 14 15 16 17 18 / 19 20 21 22 23 24 25 / 26 27 28

MARCH — S M T W T F S — 1 2 3 4 / 5 6 7 8 9 10 11 / 12 13 14 15 16 17 18 / 19 20 21 22 23 24 25 / 26 27 28 29 30 31

APRIL — S M T W T F S — 1 / 2 3 4 5 6 7 8 / 9 10 11 12 13 14 15 / 16 17 18 19 20 21 22 / 23 24 25 26 27 28 29 / 30

MAY — S M T W T F S — 1 2 3 4 5 6 / 7 8 9 10 11 12 13 / 14 15 16 17 18 19 20 / 21 22 23 24 25 26 27 / 28 29 30 31

JUNE — S M T W T F S — 1 2 3 / 4 5 6 7 8 9 10 / 11 12 13 14 15 16 17 / 18 19 20 21 22 23 24 / 25 26 27 28 29 30

JULY — S M T W T F S — 1 / 2 3 4 5 6 7 8 / 9 10 11 12 13 14 15 / 16 17 18 19 20 21 22 / 23 24 25 26 27 28 29 / 30 31

AUGUST — S M T W T F S — 1 2 3 4 5 / 6 7 8 9 10 11 12 / 13 14 15 16 17 18 19 / 20 21 22 23 24 25 26 / 27 28 29 30 31

SEPTEMBER — S M T W T F S — 1 2 / 3 4 5 6 7 8 9 / 10 11 12 13 14 15 16 / 17 18 19 20 21 22 23 / 24 25 26 27 28 29 30

OCTOBER — S M T W T F S — 1 2 3 4 5 6 7 / 8 9 10 11 12 13 14 / 15 16 17 18 19 20 21 / 22 23 24 25 26 27 28 / 29 30 31

NOVEMBER — S M T W T F S — 1 2 3 4 / 5 6 7 8 9 10 11 / 12 13 14 15 16 17 18 / 19 20 21 22 23 24 25 / 26 27 28 29 30

DECEMBER — S M T W T F S — 1 2 / 3 4 5 6 7 8 9 / 10 11 12 13 14 15 16 / 17 18 19 20 21 22 23 / 24 25 26 27 28 29 30 / 31

Geometry and Measure

Guided practice

Can you think of something you might do in each of these months?

< OCTOBER > — S M T W T F S

< NOVEMBER > — S M T W T F S

< DECEMBER > — S M T W T F S

Lesson 3: **O'clock**

• Recognise o'clock times

Let's learn

minute hand

hour hand

This clock shows 3 o'clock.

Guided practice

Write the time the clock shows.

8 o'clock

Lesson 4: **Half past**

• Recognise half past times

Let's learn

Geometry and Measure

Guided practice

Tick the clocks that show half past times.

Lesson 1: **Recognising 2D shapes**

- Recognise circles, squares, triangles and rectangles

Key words
- 2D shape
- circle
- square
- triangle
- rectangle

Let's learn

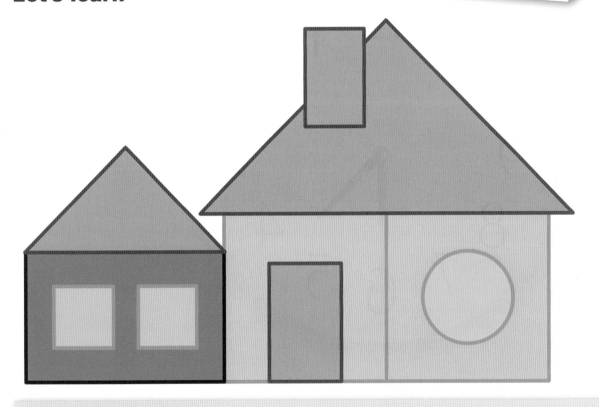

Guided practice

Colour the squares green.

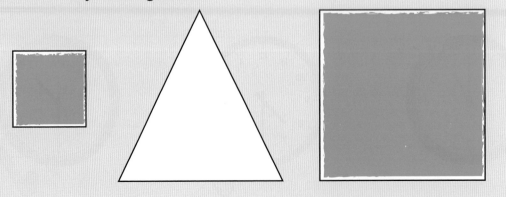

Lesson 2: **Describing 2D shapes**

- Recognise the features of circles, squares, triangles and rectangles

Let's learn

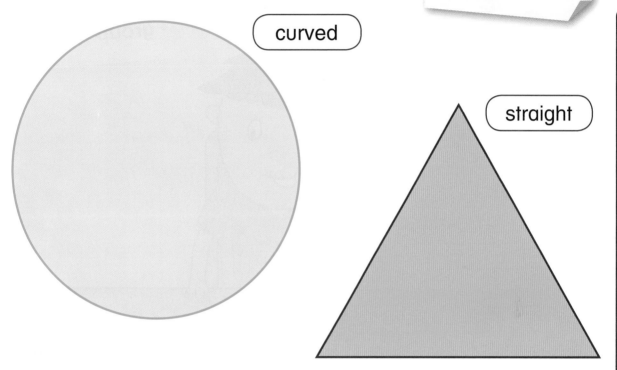

curved

straight

Geometry and Measure

Guided practice

Draw lines to join the shapes to the correct box.

curved side

4 sides

Lesson 3: **Sorting 2D shapes**

• Sort circles, squares, triangles and rectangles

Key words
• **2D shape**
• **sides**
• **straight**
• **curved**
• **sorting**
• **group**

Let's learn

Geometry and Measure

Guided practice

Tick all the shapes with three sides.

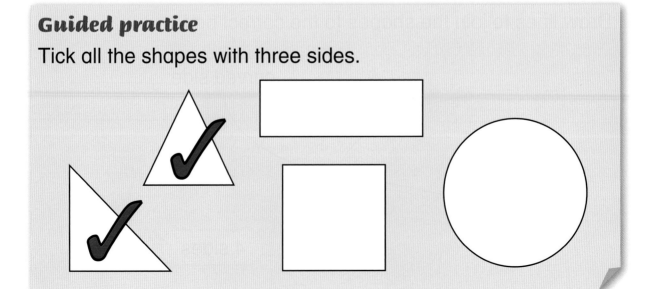

Lesson 4: **Rotating shapes**

- Identify when a rotated shape looks the same
- Describe or extend a repeating pattern

Key words
- **2D shape**
- **turn**
- **rotate**

Let's learn

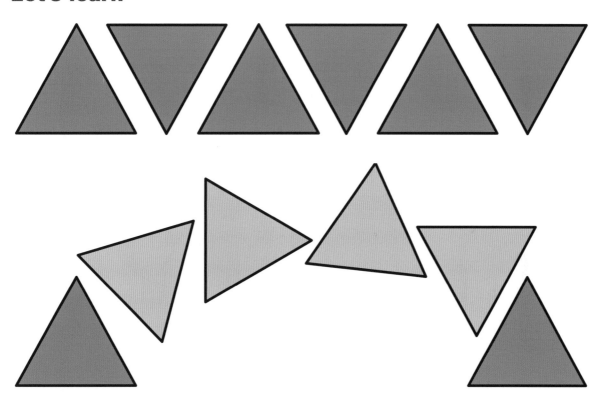

Geometry and Measure

Guided practice

Colour the shapes that will look the same when they have been rotated.

Lesson 1: **What is a 3D shape?**

- Recognise 3D shapes
- Name 3D shapes

Let's learn

Geometry and Measure

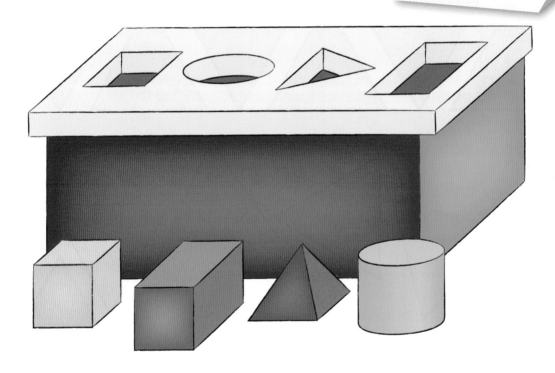

Guided practice

Join each 3D shape to its name.
Watch out for the 2D shape!

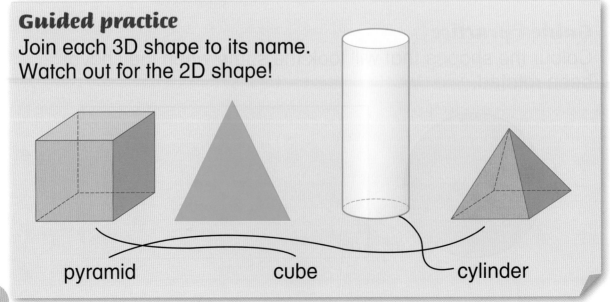

pyramid   cube   cylinder

Lesson 2: **Making models with 3D shapes**

- Make models with 3D shapes
- Identify when a rotated shape looks the same

Let's learn

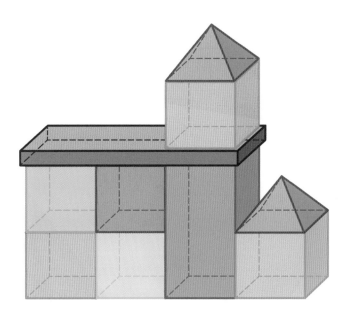

Guided practice

These three shapes were used to make the model below.

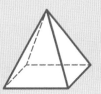

Tick the shapes that have been rotated.

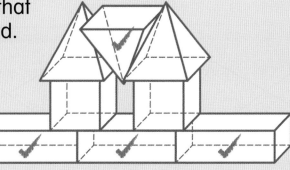

Lesson 3: **Describing 3D shapes**

Key words
- **3D shape**
- **faces**
- **edges**
- **curved**
- **flat**

- Recognise the features of cubes, cuboids, cylinders, spheres and pyramids

Geometry and Measure

Let's learn

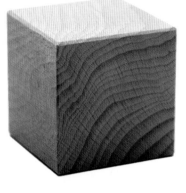

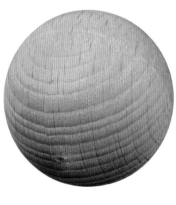
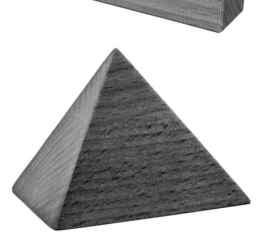

Guided practice

Draw lines to match each shape to the number of faces and the description.

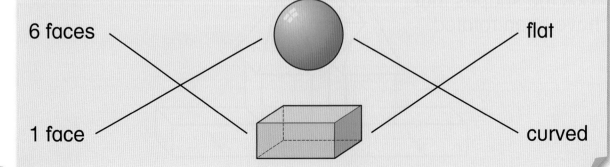

6 faces   flat

1 face   curved

Lesson 4: **Sorting shapes**

- Sort 3D shapes

Key words
- **3D shape**
- **sort**
- **faces**
- **edges**
- **curved**
- **flat**

Let's learn

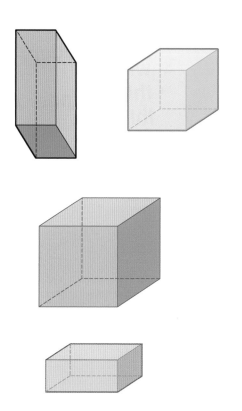
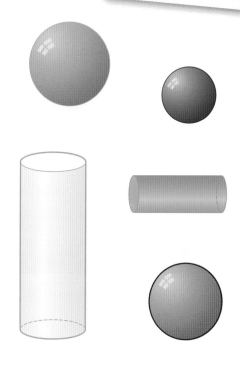

Geometry and Measure

Guided practice

Colour the shape with a curved face.

Tick the shape with 6 faces.

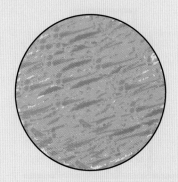
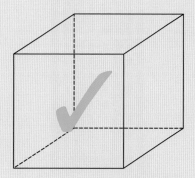

Lesson 1: **Length, height and width**

• Compare and order length, height and width

Let's learn

Guided practice
Draw a **shorter** building.

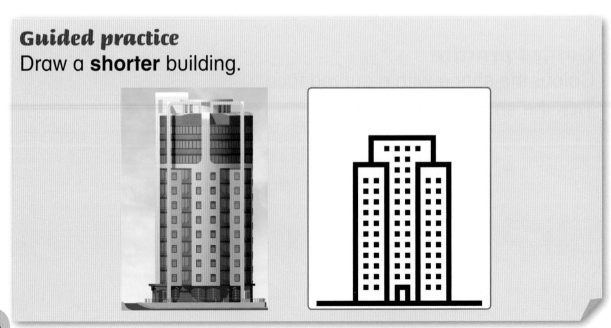

Lesson 2: **Measuring length**

• Measure the lengths of objects

Let's learn

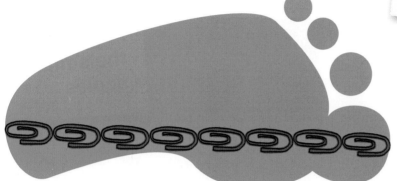

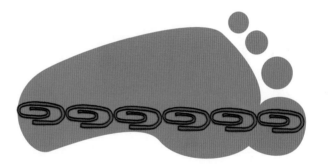

Geometry and Measure

Guided practice
Use cubes to measure the pencil case.

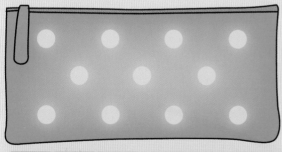

6 cubes

95

Lesson 3: **Mass**

- Compare the mass of objects

Let's learn

Key words
- measure
- mass
- heavy /
  heavier /
  heaviest
- light / lighter
  / lightest
- the same
- scales

Guided practice
Draw something **lighter**.

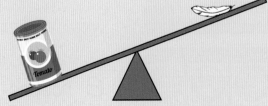

Lesson 4: **Measuring mass**

- Use everyday objects to measure mass

Key words
- **mass**
- **measure**
- **balance**
- **scales**

Let's learn

Geometry and Measure

Guided practice

Do you add or take away cubes to make the scale **balance**? Circle your answer.

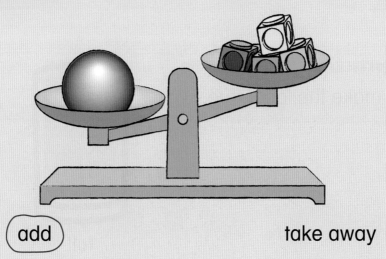

(add)   take away

Lesson 1: **Full, half full or empty?**

- Recognise when a container is full, half full or empty

Let's learn

Guided practice

Colour to make the label true.

full

Geometry and Measure

Lesson 2: **Estimating and comparing capacity**

• Estimate and compare capacity

Let's learn

Geometry and Measure

Guided practice

Order the capacities from **least** (1) to **most** (4).

| 4 | 1 | 3 | 2 |

Lesson 3: **Measuring capacity**

- Measure capacity in units of measurement that are the same

Key words
- capacity
- measure
- more
- less
- most
- least

Let's learn

Geometry and Measure

Guided practice
Use the yoghurt pots to fill the jug with water. How many yoghurt pots did you need to fill the jug?

Draw them.

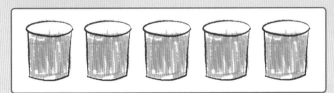

Lesson 4: **Temperature**

- Identify when we feel hot or cold
- Recognise when an object is hot or cold

Key words
- warm
- hot
- hottest
- cool
- cold
- coldest

Let's learn

Geometry and Measure

Guided practice
Colour the **hot** things red and the **cold** things blue.

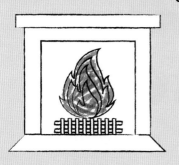

Lesson 1: **Describing direction**

- Give directions to move an object

Key words
- **up**
- **down**
- **forwards**
- **backwards**
- **around**

Let's learn

Guided practice

Where is Ted going?
Circle the correct word.

up   around   backwards

Lesson 2: **Left and right**

• Know which way is left and which way is right

Key words
• **left**
• **right**
• **direction**
• **turn**

Let's learn

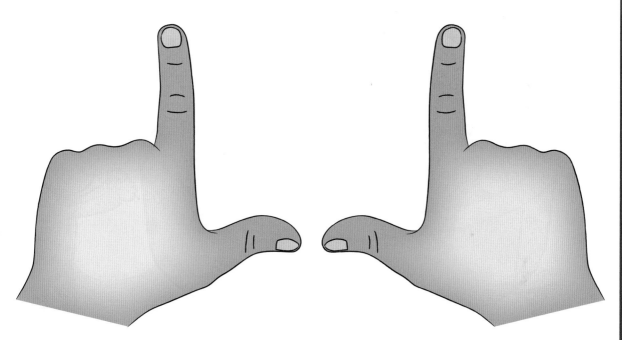

Geometry and Measure

Guided practice
Which way is the sign pointing?

<u>left</u>

Lesson 3: **Describing position**

• Describe an object's position

Let's learn

> **Key words**
> • **on**
> • **inside**
> • **outside**
> • **under**
> • **over**
> • **behind**
> • **next to**
> • **beside**

Guided practice
Where is the cat?

on

outside

behind

under

inside

Lesson 4: **Following directions**

• Follow directions

Let's learn

Walk forwards...
now turn left.

Key words
• **forwards**
• **backwards**
• **left**
• **right**
• **up**
• **down**
• **around**

Geometry and Measure

Guided practice
Draw a line to show the ball's movements:
• **up** the **right** side of the pole
• **over** the **top** of the net
• **down** the **left** side of the pole.

Lesson 1: **Sorting data**

• Sort data

Key words
• statistics
• data
• sort
• the same
• different

Let's learn

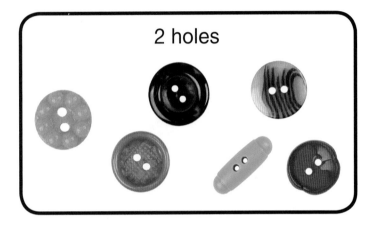

2 holes

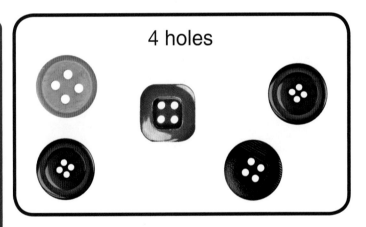

4 holes

Statistics and Probability

Guided practice

Draw lines to sort the buttons.

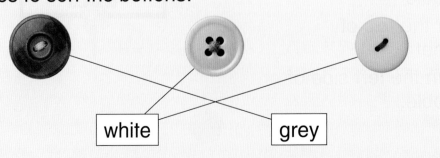

white   grey

Lesson 2: **Collecting data**

• Collect data to answer a question

Let's learn

Guided practice

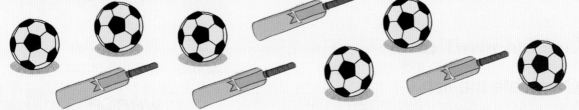

Favourite sport

How many people like football best? 6

How many people like cricket best? 4

What is the most popular sport? <u>*football*</u>

Lesson 3: **Using lists**

- Organise data in a list

Let's learn

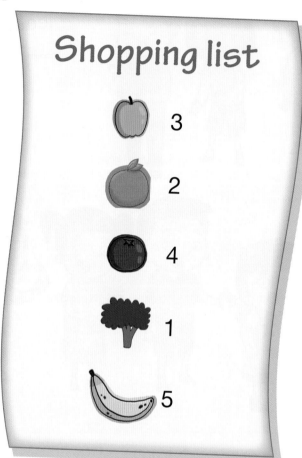

Shopping list

Guided practice

Complete the list.

Sweets

Lesson 4: **Using tables**

• Organise data in a table

Let's learn

Bike	Number sold
	5
	1
	4
	2

Guided practice
Complete the table.

Pet	Total
	4
	1
	2

Statistics and Probability

Lesson 1: **Venn diagrams**

- Use a Venn diagram to sort information

Key words
- data
- Venn diagram
- sorting rule

Let's learn

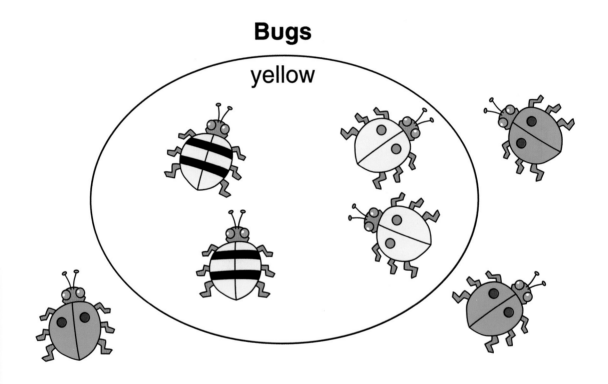

Bugs

yellow

Guided practice

Complete the Venn diagram.

Food

Fruit and vegetables

Statistics and Probability

Lesson 2: **Carroll diagrams**

• Use a Carroll diagram to sort information

Let's learn

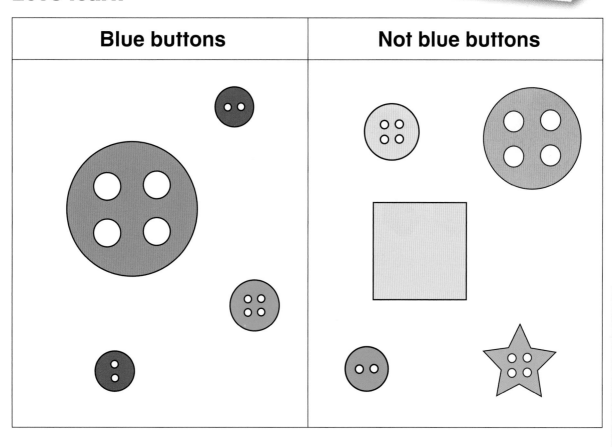

Blue buttons	Not blue buttons

Guided practice

From the beach	Not from the beach

Statistics and Probability

Lesson 3: **Pictograms**

• Use a pictogram to show information

Let's learn

Class 1's shoes

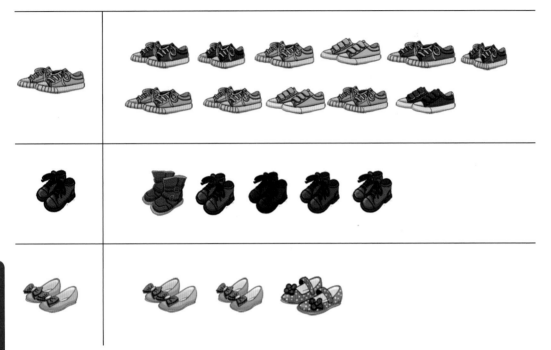

Guided practice
Write the totals.

Favourite toys							Total
ball	⚽ ⚽ ⚽ ⚽ ⚽ ⚽ ⚽						7
car	🚗 🚗 🚗 🚗 🚗						5
doll	👧 👧 👧 👧 👧 👧						6

Lesson 4: **Block graphs**

• Use a block graph to show information

Key words
• block graph
• label

Let's learn

Favourite yoghurt

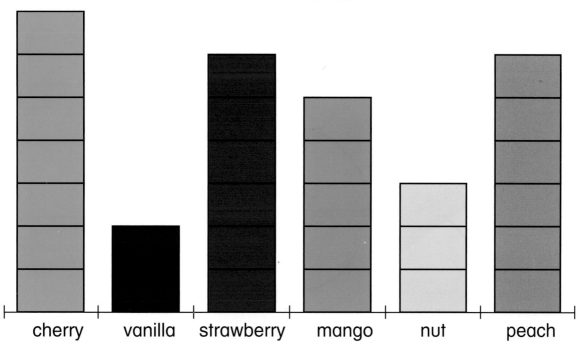

| cherry | vanilla | strawberry | mango | nut | peach |

Guided practice
Complete the block graph.
Then write a statement
about the block graph.

🌧	2
☀	4
☁	1

<u>There were more sunny
days than wet days.</u>

Weather last week

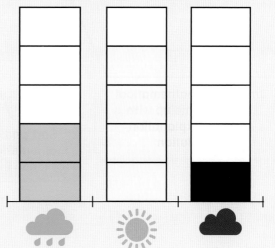

The Thinking and Working Mathematically Star

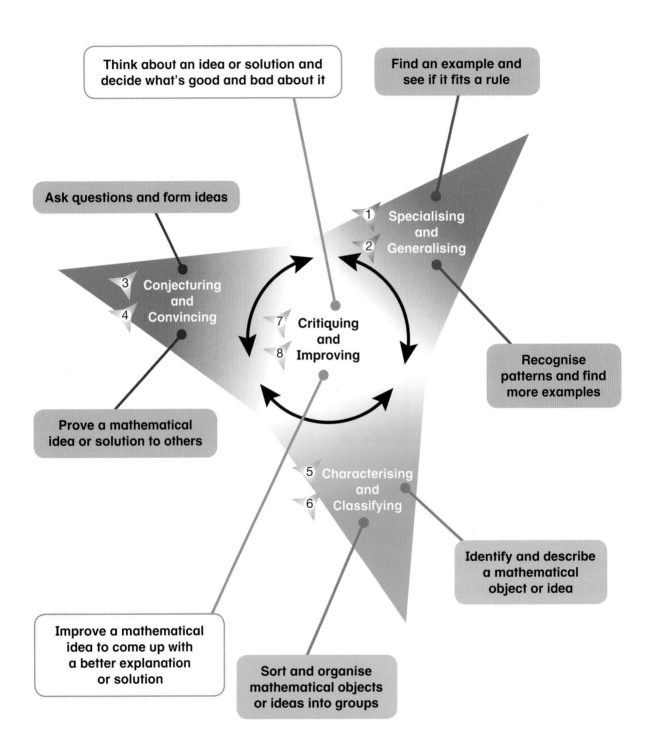

Think about an idea or solution and decide what's good and bad about it

Find an example and see if it fits a rule

Ask questions and form ideas

1 2 **Specialising and Generalising**

3 4 **Conjecturing and Convincing**

7 8 **Critiquing and Improving**

Recognise patterns and find more examples

Prove a mathematical idea or solution to others

5 6 **Characterising and Classifying**

Identify and describe a mathematical object or idea

Improve a mathematical idea to come up with a better explanation or solution

Sort and organise mathematical objects or ideas into groups

Acknowledgements

Photo acknowledgements

Every effort has been made to trace copyright holders. Any omission will be rectified at the first opportunity.

p6b Madlen/Shutterstock; p8t MIKHAIL GRACHIKOV/Shutterstock; p9t Apnea/Shutterstock; p10b Olgers/Shutterstock; p15t Olga Tagaeva/Shutterstock; p15b Magicleaf/Shutterstock; p16tl Rodynchenko/Shutterstock; p16tc CWIS/Shutterstock; p16tr Bergamont/Shutterstock; p17t Mcimage/Shutterstock; p17b Vladimir Volodin/Shutterstock; p22bl Magicleaf/Shutterstock; p22br Olga Tagaeva/Shutterstock; p23b Rvector/Shutterstock; p24tl Gelpi/Shutterstock; p24tcl Africa Studio/Shutterstock; p24tcr Slatan/Shutterstock; p24tr Cheese78/Shutterstock; 24cl Gelpi/Shutterstock; p24c Novikov/Shutterstock; p24cr Piyawat Nandeenopparit/Shutterstock; p38b Robuart/Shutterstock; p44cl Cristina Ionescu/Shutterstock; p44cr Suradech Prapairat/Shutterstock; p58tl Irin-k/Shutterstock; p58tr PHOTOCREO Michal Bednarek/Shutterstock; p58cl Butterfly Hunter/Shutterstock; p58cr Kuttelvaserova Stuchelova/Shutterstock; p62t Mubus7/Shutterstock; p62bl Miniyama/Shutterstock; p66b Olgers/Shutterstock; p69t Sergey Novikov/Shutterstock; p74t Maks Narodenko/Shutterstock; p79r Sweet Art/Shutterstock; p80t Theskaman306/Shutterstock; p84b Dmitry Zimin/Shutterstock; p85t Szefei/Shutterstock; p85b ColinCramm/Shutterstock; p92t Laborant/Shutterstock; p92cr Olga Vorontsva/Shutterstock; p94t 7th Son Studio/Shutterstock; p94bl Anton Gvozdikov/Shutterstock; p94br Howcolour/Shutterstock; p96t Olesia Bilkel/Shutterstock; p96b Picture Store/Shutterstock; p98t Yuliyan Velchev/Shutterstock; p98b NikWB/Shutterstock; p99t Isaree/Shutterstock; p99bl Doomu/Shutterstock; p99bcl Olga Popova/Shutterstock; p99bcr Rawf8/Shutterstock; p99br DenisProduction.com/Shutterstock; p100tl Nito/Shutterstock; p100tr Volab/Shutterstock; p102t Marlon Lopez MMG1 Design/Shutterstock; p103b Peter Gudella/Shutterstock; p104b Vitaly Titov/Shutterstock; p105t IMAGE LAGOON/Shutterstock; p105bl Alfonso de Tomas/Shutterstock; p105br Vectoric/Shutterstock; p106b Donatas1205/Shutterstock; p110bl Xpixel/Shutterstock; p110bccl LarysArty/Shutterstock; p110bcl Art Alex/Shutterstock; p110bc Sunnydream/Shutterstock; p110bcr N.Petrosyan/Shutterstock; p110bccr Visual Generation/Shutterstock; p110br IgorijArt/Shutterstock; p111 Tahahnka/Shutterstock; p111blc Ivaschenko Roman/Shutterstock; p111blrt Kletr/Shutterstock; p111bbl Creatsy/Shutterstock; p111blrb Ed2806/Shutterstock; p111bct Frescomovie/Shutterstock; p111bcb Sabelskaya/Shutterstock; p111bcr Egor Shilov/Shutterstock; p111btr BeatWalk/Shutterstock; p111bbr Christopher Hall/Shutterstock; p113bl Notbad/Shutterstock; p111bc WEB-DESIGN/Shutterstock; p111br AF Studio/Shutterstock.